THE NATURE OF LIONS

Text by Eric S. Grace

Photographs by Art Wolfe

The Nature of LIONS

Social Cats of the Savannas

Firefly Books

To Jonathon "Leo" Neville, Jasper, and Rupert

Published by Firefly Books (U.S.) Inc. 2001

PAGES IV–V
Lions survey the savanna landscape of their home in Kenya.

First Printing

U.S. CATALOGING-IN-PUBLICATION DATA
(Library of Congress Standards)

Grace, Eric S.
The nature of lions : social cats of the savannas / text by Eric S. Grace; photographs by Art Wolfe. –1st ed.
112 p. ; col. photos. : cm.
Includes index.
Summary: The natural history of the lions of Africa's savanna habitats.
ISBN 1-55297-542-8
1. Lions—Behavior—Africa. 2. Lions—Habitat—Africa. 3. Lions—Africa, Central. I. Wolfe, Art. II. Title.
599.757'96 21 2001 CIP

Originated by Greystone Books in Canada and published simultaneously in the United States by Firefly Books.

Published in the United States in 2001 by
Firefly Books (U.S.) Inc.
P.O. Box 1338, Ellicott Station
Buffalo, New York 14205

Text and jacket design by Tanya Lloyd and Val Speidel
Editing by Nancy Flight
Jacket photographs by Art Wolfe
Printed and bound in Hong Kong by C & C Offset Printing Co., Ltd.

CONTENTS

PREFACE

The lion sits sphinxlike in the shade of an acacia tree. Front paws aligned, face serene, he does not acknowledge that a truckload of humans has stopped 20 meters (65 feet) away. The truck's engine is turned off, and the passengers sweat under the late afternoon sun. A fly settles on the broad slope of the lion's nose, then dances up toward the dark, moist corner of one eye. I marvel at its impertinence.

Wildebeests grunt behind us. Weaver birds flit overhead and then settle, chattering, among the thorny branches of the acacia tree. There is a low murmur among us humans. The click and whir of cameras irritate me but do not concern the lion.

The lion has seduced us. He seems embraceable. He is a giant pussy cat. One of the people in the truck must think so, for he jokingly puts a leg over the side of the truck as if to jump down and stroll over to pet the golden fur.

In the 1941 movie version of *Dr. Jekyll and Mr. Hyde,* Spencer Tracy artfully changes his face and body from benign to sinister with the subtlest of maneuvers. Cued by the movement from the truck, the lion (he isn't acting), like Tracy, is transfigured in an instant, with only a very slight turn of his head. Eyes that had been soft and unfocused become hard yellow beacons. Muscles that had yielded to gravity tense into rigid contours. The pussy cat is a predator. No more MGM mascot, he stars in horror flicks featuring tourists dragged from tents in the dark, cold night by unforgiving jaws.

* * *

FACING PAGE
Serene and aloof in repose, lions can quickly transform themselves into formidable predators.

THE LION EMBODIES SUCH A POWERFUL COMBINATION OF BEAUTY AND BEAST, SUCH DRAMATIC GRACE, THAT IT HAS CAPTURED THE HUMAN IMAGINATION SINCE PRE-HISTORIC TIMES.

What are lions? Surely they are more than my encyclopedia's succinct and rather droll introduction to them as "large roaring cats." Yet again, on reflection, these three words assume in my mind the subtle poetry of a haiku. They are accurate. Certain.

The lion embodies such a powerful combination of beauty and beast, such dramatic grace, that it has captured the human imagination since prehistoric times. In those days, our ancestors elbowed lions from the caves in which both sought shelter, then etched images of these beasts on the cave walls. People and lions are longtime rivals for space, and our shared and bloody history gives human and cat a mutual sense of respect and fear. Inevitably, we endow these carnivores with attributes of our own, giving them roles in our storybooks and transforming them into symbols.

With what eyes must I see lions today, when most people live in cities, our technology dominates the planet, and there are more humans and fewer wild carnivores than at any previous time in our history? In these pages, I set out to discover the nature of lions, knowing at the start that this is an impossible task.

There is the tourist's lion, usually asleep, providing photo opportunities on vacations. The circus trainer's lion, with its echoes of the ancient power struggles between our species. The zookeeper's lion, whose snarls at feeding time evoke the atavistic excitement of nights by a campfire on the African plains. The symbolic lion, far more plentiful than real lions and seen throughout the world in statues, paintings, tapestries, stained glass windows, storybooks, flags, coins, advertisements, door knockers, and many other places. There is the hunter's lion, the livestock farmer's lion, the zoologist's lion, and the writer's lion.

Lions today seem quintessentially African. They parade before our mind's eye in scenes populated by gazelles and zebras, elephants and giraffes. They loll in the equatorial sun on slides from safaris, hunt wildebeests in TV documentaries, form stylish tableaus against dusk-red horizons in the photo spreads of magazines, and lie with closed eyes in black and white at the feet of hunters posed for posterity inside musty books. Yet these typically African inhabitants were part of the landscape over much wider areas of the world until not so long ago.

Chapter 1 of this book traces the evolution of lions and the characteristics that make them unique among cats. Chapter 2 describes their social structure and life history, as well as the web of climate, land, plants, and other animals that shape their lives. The final chapter reviews some historical connections between lions and people and looks at the troubled future facing lions.

Chapter One
LARGE ROARING CATS

Humans and lions both owe their lives to two planetary traumas, one originating in outer space, the other coming from deep below the Earth's surface. The extraterrestrial event occurred some 65 million years ago, when a large meteorite struck the planet. We still have plenty of evidence from the fallout of that encounter to recreate the story. With the force of a bomb, Earth and space rock collided, hurling a craterful of debris into the atmosphere. It was not the collision itself but its aftermath that had the greatest impact on life. Clouds of dust, together with smoke and soot from burning forests, circled the globe, and the world grew cold and dark. An estimated one-quarter of all land vegetation died. Soon many large herbivores and their predators also disappeared.

The debate continues as to whether this event alone was the cause of mass extinctions near the end of the Cretaceous period or whether it was only one of a sequence of related disasters. Mounting recent evidence points overwhelmingly to the conclusion that large meteorite impacts have caused major disruptions to life on Earth. The best-known victims of the Cretaceous impact were the large dinosaurs that had dominated the land for over 120 million years and were never to be seen alive again.

Nature has no favorites. The dramatic extinction of the dinosaurs created new opportunities for animals that survived the cataclysm. Among them were small, active mammals that were able to find food from a wide range of sources. Without competition from the dinosaurs, the mammals rapidly diversified to occupy ecological niches left vacant by the vanished species. One line of mammalian development eventually led to lions, another to human beings.

FACING PAGE

Carnivorous mammals such as the lion can trace their ancestry back millions of years to small, insect-eating animals that lived on Earth when dinosaurs dominated the land.

FACING PAGE
Muzzle raised and eyes closed, a lioness rakes her claws down the trunk of a tree to help keep them sharp.

When the Age of Mammals replaced the Age of Reptiles, mammals had already been around for a very long time. The first identifiable members of this group appear in fossils more than 200 million years old. At that time, Earth's land masses were joined in a single large continent. Over the following millennia, while dinosaurs dominated the landscape, upwellings of molten rock from the planet's interior broke the supercontinent apart, first into two and then into a half-dozen or so titanic platforms. As the 100-kilometer (60-mile)-thick plates moved continents on their backs like boxes on a conveyor belt, they also carried their living passengers separate ways.

The breakup of continents and the associated building of mountain ranges isolated populations of animals in different environments, accelerating the evolution of new characteristics. Some mammals remained small and nocturnal, with versatile habits—the rodents. Some occupied open grasslands and became large and swift—the hoofed grazers. Some took to the trees—the primates. From teeth and bones bequeathed to us by long-dead species, we can trace the roots of today's diverse mammal groups to this long period of continental realignment.

THE EVOLUTION OF THE LION

The lion is only one among hundreds of modern examples of the branch of mammals that produced carnivores. "Carnivores" here refers not to meat eating animals in general but specifically to the order of mammals called Carnivora. This order includes various families of cats, dogs, weasels, and hyenas, as well as some omnivores, such as bears and raccoons. It also includes a few animals that are mainly plant eaters, such as the giant panda and an equally delightful creature called the binturong—a nocturnal, fruit-eating inhabitant of forests in southeast Asia.

All carnivores have a relatively simple stomach and digestive system, since a diet of animal flesh is generally easier to digest than are the tougher tissues of plant matter. Carnivores also share many details of their teeth, jaws, and skull structure—an essential field of study for taxonomists. The key to the carnivores' success is a front end designed for shearing through meat and, with it, the brainpower, behavior, and senses needed to back up their challenging way of life.

The hallmark of true carnivores is two pairs of bladelike carnassial teeth, located about halfway between the front of the jaw and the jaw joint. The upper and lower carnassials on each side work together like a pair of scissors, allowing carnivores to slice off strips of flesh. The oldest fossil skulls bearing carnassial teeth are about seventy million years old. These skulls belonged to small, weasel-like creatures that appeared on the scene just before the dinosaurs became extinct.

Shortly after the carnivore line was launched, the stock split into two main branches: the doglike (superfamily Canoidea) and the catlike (superfamily Feloidea). The doglike line, including wolves, foxes, bears, weasels, and

FACING PAGE
Lions obtain much of the moisture they need from their meals and can survive even in dry desert areas if there is enough prey to live on.

THE EVOLVING FORMS OF THE CATLIKE CARNIVORES WERE SHAPED IN PART BY THE EVOLVING DEFENSES OF THEIR QUARRY.

raccoons, kept a tooth pattern much like that of their ancestors. They still have a full set of teeth that can crush bones and chew insects and plants as well as shear flesh, giving this group a widely varied diet. Cats, however, became more and more specialized for the meat-eating life. They lost their crushing and grinding molars and are unable to chew their food. The most committedly carnivorous of the carnivores, cats can only tear off and swallow chunks of meat.

Feline ancestors with the features of modern cats first appeared six million years ago—about the same time as humanlike apes. The evolving forms of the catlike carnivores were shaped in part by the evolving defenses of their quarry. Better hunters produce wilier prey, wilier prey produce more efficient hunters, and so the endless game of cat and mouse goes on. Within the cat family (Felidae), ancestors with the slender, leggy build to outrun their prey on open ground produced the cheetah (subfamily Acinonyxinae). Smaller and more compact ancestors hunted birds and small mammals in forests or on rocky terrain, giving rise to various species of wild cats, servals, ocelots, lynxes, and their relatives (all subfamily Felinae). The largest of this group is the puma. Bigger felines turned their attention to deer, boar, and other substantial prey. They are represented today by only five species: snow leopards, leopards, jaguars, tigers, and lions (all subfamily Pantherinae).

At the start of the Pleistocene era, some two million years ago, herds of longhorned bison, giant elk, mammoths, woolly rhinoceroses, and other large, thick-skinned grazers roamed the plains, ever alert for powerful hunters such as saber-toothed cats and giant cave lions. Adapted to life on the cold, open steppes and prairies, cave lions were widespread across Eurasia from England to Siberia. Together with other animals, including early humans, they spread to Alaska across the grassy Bering isthmus, which was

TO PUT THEIR EVOLUTIONARY HISTORY IN PERSPECTIVE, ALL OF TODAY'S LARGE CATS ARE, LIKE *HOMO SAPIENS*, NEWCOMERS ON THE GEOLOGICAL TIME SCALE.

exposed during periods of glaciation when sea levels were lower. American lions eventually ranged as far south as Peru. We know what they looked like from the many well-preserved specimens that have been excavated from the La Brea tar pits in California. Although nearly 25 percent larger than modern African lions, they were very similar to today's lions in the structure of their teeth and skeletons and probably in their behavior.

As the climate warmed at the end of the last ice age, many large mammals became extinct. Other species thrived or moved into new environments. Ancestral lions followed their large prey from Eurasia into Africa by traveling up the Nile River valley and then dispersed throughout the African continent. Taxonomists distinguish six subspecies of the lion (*Panthera leo*) in Africa, of which two—the Barbary lion of the north coast and the Cape lion of the south coast—are now extinct. The Barbary lion is the type that appeared in Roman circuses. The four remaining African subspecies are the Angolan, the Senegalese, the Masai, and the Transvaal lions. The Asiatic lion (*Panthera leo persica*) is the only subspecies now living in the wild outside of Africa. It barely survives as a tiny population in the Gir Forest of western India. Genetic comparisons have led scientists to conclude that the Asiatic lion split from the African population about 100,000 years ago. Subspecies differ in small physical details such as size, color, mane growth, and skin fold along the belly. The genetic differences among these subspecies are minor, being even less than the differences among human racial groups.

To put their evolutionary history in perspective, all of today's large cats are, like *Homo sapiens,* newcomers on the geological time scale. Lions, tigers, and leopards separated into different species so recently, and are so close genetically, that they can be cross-bred in captivity.

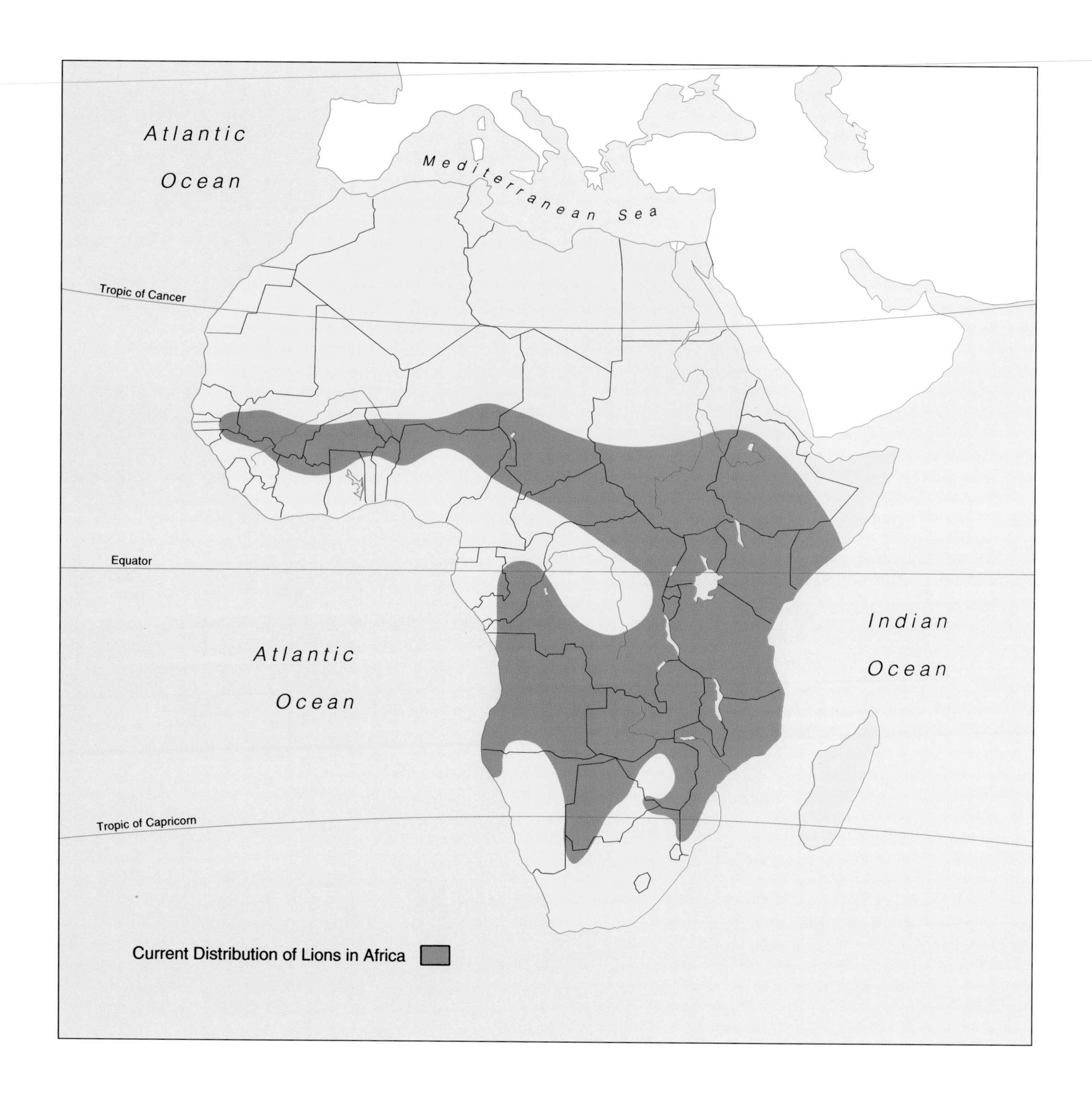
Atlantic
Ocean
Mediterranean Sea
Tropic of Cancer
Equator
Atlantic
Ocean
Indian
Ocean
Tropic of Capricorn
Current Distribution of Lions in Africa

IN THE SKIN OF A LION

Everything about a lion's body says "hunter." The supple spine enables the lion to press its belly close to the ground when it is stalking and then to arch into a bow to catapult itself toward its prey. Powerful leg muscles launch the hunter's leap: from standing position, a lion can jump over a fence the height of two men or spring forward a distance of 12 meters (40 feet). A running lion can reach bursts of 64 kilometers per hour (40 miles per hour), so if one is charging toward you from a distance of 100 meters (300 feet), you have less than six seconds to escape.

A surprise rush is crucial to the lion's hunting success. Lions don't have much stamina for running, and their typical prey—zebra, wildebeests, impala, and gazelles—can run faster than them. With a flying leap onto the back of its fleeing meal, the lion demonstrates another specialization. Smaller and more primitive carnivores, such as polecats and civets, use only their jaws to catch and kill, and this generally limits their victims to animals not much larger than their heads. But the larger cats, by using their claws and forelimbs to seize prey, can tackle much larger game. Their claws, built-in grappling hooks, are unsheathed only when needed and are protected from blunting the rest of the time inside tough, padded paws. Clinging to the tough hide of its prey, the lion uses its weight to bring the animal to a halt or topple it onto its side.

The lethal focus of the lion's jaws is on the neck or muzzle of its captive, but the actual method of killing varies. Smaller victims lose their lives to the penetrating wounds of a bite from the lion's daggerlike teeth. Larger prey more commonly die by being strangled or suffocated. Using the powerful jaw muscles anchored to its massive skull, a lion clamps its maw around a

FACING PAGE
This map shows the approximate distribution of lions in Africa today. Within the boundaries of each country, lions are found mainly in shrinking areas of wilderness and in national parks. In past centuries, lions were found throughout most of the continent.

zebra's throat to squeeze the windpipe shut or engulfs the unfortunate animal's nose and mouth to stop its breath.

Before the claws and jaws come into play, before the run and leap, a lion relies on its sense organs. As with people, the lion's chief sense is sight. It hunts almost entirely by watching its quarry, raising its head above the vegetation if necessary to keep the prey in view. The large, forward-pointing eyes give the lion stereoscopic, binocular vision like our own, providing accurate information about distance. A lion's ability to accurately judge the distance to its prey can determine whether its final killing rush will be successful or not.

Lions hunt when opportunity presents itself but most frequently search for prey under cover of darkness. Like other nocturnal animals, they have a reflective layer at the back of their eyes that amplifies the light entering the eyeball and helps them see more clearly in dim conditions. It is this reflective layer that makes an animal's eyes shine in the dark. The shape of the pupil in lions and other big cats (the Pantherinae) is oval to round, like ours, not the vertical slit of the house cat and other small cats. Differences in pupil shapes among various predators likely result from a combination of adaptations to such things as light levels when the predators are most active, the depth of field, and the size of their prey.

Ears are next in importance to a lion, especially at night. A splash in a creek, a rustle of dry grass, the cries of hyenas, and the roars of other lions—all are snatches of stories that tell of opportunity or danger. The sense of smell trails behind vision and hearing, serving a lion only where potential prey cannot be seen or heard, as when it is sniffing down a warthog hole or an aardvark burrow. Detecting scents is more important in social interactions, such as marking territories and identifying females in heat.

FACING PAGE
Among the most fearsome weapons in the lion's arsenal is a set of teeth shaped to cut and tear flesh.

Just behind its nose is the lion's fourth sense. From either side of its muzzle sprouts a straggly bunch of fine white whiskers, each growing from a small spot of dark fur. The wiry whiskers are connected at their roots to nerves and are sensitive to touch.

The body of a lion is second in size only to that of the Siberian tiger in the cat family. Adult male lions weigh between 160 and 200 kilograms (between 350 and 440 pounds) and stand about 1 meter (just over 3 feet) at the shoulder. Females are about three-quarters this size. Despite their bulk, lions are not always easy to spot. Their coat color blends with the semiarid landscape, especially when they are not moving. Paler hair on their undersides neutralizes telltale shading when they stand in the sun, and at night, by starlight, lions appear uniformly gray and ghostlike as they move soundlessly on soft pads.

In contrast to all other cats, there is a marked difference in the appearance of male and female lions. Not only are males larger, they bear their species' hallmark feature, the mane. Like men's beards, this regal mantle varies greatly in color and fullness from individual to individual. It is pale yellow on this lion, dark brown on that one, and black on the other. One lion sports only a modest ruff on top of his head and behind his ears, while another has a heavy growth that frames his face, covers his neck and chest, and extends well over his shoulders.

FACING PAGE
This playful young cub will have to wait several years before he has the magnificent mane of a full-grown adult.

Why do lions have such a conspicuous feature, unique among cats? Some sources suggest that the mane protects a lion's throat during his rare but savage fights with other males. This argument isn't very persuasive, however, since lionesses and maneless cat species also fight. A more likely explanation is that the mane is an advertisement of male potency. Like the peacock's tail, which is useless for anything but display, a mane signals genetic vigor. The

A PAIR OF LIONS LICKING EACH OTHER'S FACE IS A PICTURE OF AFFECTION, BUT WHEN A LION ACTS FRIENDLY, IT'S FOR A REASON.

owner of the biggest and best mane eloquently telegraphs his strength and superiority to males and females alike, discouraging the former and attracting the latter. Symbolic displays are widespread among many social species of mammals, birds, fish, and reptiles. They allow individuals to compete with each other by proxy, substituting bigger shapes, brighter colors, or louder sounds for the risk and effort of actual combat.

COMMUNICATION

It is difficult for cat lovers not to feel flattered by the head-rubbing display that greets them on reuniting with their pet. It is the standard feline welcome. When returning to its pride after an absence, a lion brushes its cheeks against those of each pride member in turn. This act looks so sensual that it makes one think that noserubbing should be more widely adopted as a salutation among humans. On the practical side, this behavior may spread oily scents from glands on the face, ensuring that all pride members share the same smell. The mutual massage may be prolonged and, if both lions are standing, may include rubbing the sides of their bodies together. Similar cheek-rubbing behavior is directed by estrous females toward males and by hungry cubs toward adults.

Like domestic cats, lions spend much time running their tongues over fur (either their own or a companion's). Most of the upper surface of the lion's tongue is covered with tiny, sharp, backward-pointing hooks, which function as a comb for grooming and as a rasp for scraping bits of meat off bones. A pair of lions licking each other's face is a picture of affection, but when a lion acts friendly, it's for a reason. Besides the immediate and practical benefits of the "I'll scratch your back if you'll scratch mine" variety, the tactile

FACING PAGE
The lion's rough tongue is used both for grooming fur and for rasping scraps of meat from bones.

rituals of bonding reinforce the mutual cooperation on which successful breeding and hunting depend.

There is little hostility within a pride of lions—usually a growl, a snarl, or an intimidating expression is enough to convey warning, bluff, or threat. The intensity of this repertoire depends on whether it is addressed to an irritating cub disturbing a siesta or to a serious rival for a wildebeest leg. The hostile face includes ears flattened defensively against the sides of the head, lips drawn back from the teeth, a wrinkled nose, and narrowed eyes.

Lions pay close attention to the body language of other pride members and can tell another lion's disposition from its walk. An aimless amble means there is little to worry about; a walk or trot in a straight line means business afoot. A male with his head held high is asserting dominance. A lion that hurries nervously keeping its head low is trying to avoid detection. An aggressive lion holds a rigid posture with its ears forward and eyes wide open and staring. A lioness catching sight of prey can transform her mood swiftly from relaxed and drowsy indolence to focused intensity, and her alert posture spreads through the group.

Lions use their voices to meow (as cubs), hiss, snarl, yowl, grunt, and roar. When close to others, adult lions may sigh or puff, and a lioness in heat rumbles. Unlike domestic cats, lions do not purr.

"Roaring cats" is not, as it seems, a whimsical description but a taxonomic category, distinguishing the subfamily of big cats (Pantherinae) from the subfamilies that are "purring cats" (Acinonyxinae and Felinae). The ability of lions, tigers, jaguars, and leopards to roar depends on the structure of a U-shaped bone called the hyoid, which anchors and controls the motion of the tongue. In other cats, such as lynx, wild cats, and ocelots, the hyoid is

FACING PAGE
Adult male lions in a pride roar to warn away strangers.

hard and bony. In big cats the hyoid is more elastic, giving the larynx greater freedom of movement, which is needed to make roaring sounds.

So much for how lions roar. But why do they roar? Adults of both sexes can produce this awesome sound while walking, lying, or standing. They roar most frequently at night, suggesting that one function of roaring is to keep members of the group in contact while they are scattered and out of sight of one another. Individuals can recognize each other by their roars. As with the group howling of wolves, one roar may set off others until the night air vibrates with the chorus. Roaring is also territorial. Prides typically hunt within a particular range or territory, the size of which depends on the type of terrain and abundance of prey. Roars may carry for 3 kilometers (2 miles) or more, and neighboring males roar back and forth to announce their position and strength, thus helping avoid surprise encounters with rivals. Roars also release tension, and lions sometimes have short, intense roaring sessions after a kill or conflict.

LIFE IN A PRIDE

Other cats are large, other cats roar, but lions are the only cats that are truly social. Their relatives, including tigers, leopards, jaguars, and cheetahs, typically lead more solitary lives. So why do lions like company?

The complex social life of lions was something of a puzzle to people until the 1960s, when long-term studies helped sort out the ways in which lions organize themselves. In the early 1900s, for example, many observers believed that the king of beasts took his mate and settled down with her to a faithful family life. Since the best-maned lions attracted hunters as well as lionesses, it might well have been true that a king's liaison usually lasted until his death.

Facing page
Related females and their cubs form the core of a lion pride.

Later in the century, after national parks were established in Africa, many park wardens observed lions as they came and went through their area. These random encounters often left murky impressions of the big cats' private lives. Did pride members stay together for life? Did they have territories? Did they migrate? One observer in the 1950s concluded that "the probability is that there is no definite habit or custom in the matter at all."

The pioneering scientific study of lions in their natural environment occurred in the 1960s. It was carried out in Serengeti National Park, Tanzania, by George Schaller, the field biologist who had earlier studied mountain gorillas in central Africa and tigers in India. Like all great fieldworkers, Schaller combined patience, stamina, and the keen senses of a trained observer. From June 1966 until September 1969, Schaller tracked lions living in and near the park, sometimes following groups or single lions continuously for forty-eight hours or more.

By learning to recognize hundreds of animals as individuals, Schaller eventually found out that most lions belong to a pride, to which they have close ties of loyalty. The difficulty of discovering this pattern lay in the fact that members of large prides were rarely all together and could sometimes be scattered as far as 16 kilometers (10 miles) apart. Pride males travel much more than females, patrolling their territory for signs of other males and marking their boundaries with urine. Perhaps by way of compensation, they share in the pride's meals while playing little part in the hunts. Females generally move only to hunt or to find water or shade. They travel on average about 5 kilometers (3 miles) in a night, the direction depending on the location of their prey. A pride has no fixed resting place, except when females are confined to a den with small cubs.

The core of the pride consists of related adult females and their cubs. There may be anywhere from two to twenty lionesses in a pride, depending on the abundance of prey in the region and the ease or difficulty of the terrain for hunting. From one to five adult males join a pride for breeding, but their tenure typically lasts for only two to four years before they are evicted or killed by other males. When not part of a pride, male lions are nomads, wandering over large areas alone or with companions.

A pride is a more-or-less harmonious group of equals. There is no leader or dominant animal, no pecking order based on age or size—only an easygoing camaraderie among sisters, cousins, aunts, mothers, and daughters. Prides are very stable and might occupy the same area for many decades, their identities maintained through generations from grandmothers to granddaughters. The males in a pride, too, are usually closely related to one another (but born in another pride) and cooperate rather than compete with each other. Although they fight over females, the fights are not serious. Any male in a pride may mate with any female, and females may also choose one male over another.

The genetic relationships in a pride are a key to why lions are so sociable. The name of the game is evolution, and the goal of the game is survival. There are many strategies by which living things pass on their inheritance to new generations. In the simplest model of survival of the fittest, every individual is out for him- or herself, competing with every other individual to eat and reproduce. But if instead of focusing on individuals we think of the survival of genes—the idea of "the selfish gene"—then it also makes sense for organisms to help their relatives as well as themselves.

This idea, known to biologists as the theory of kin selection, explains why many social animals act in ways that appear to be altruistic. For example, a

A PRIDE IS A MORE-OR-LESS HARMONIOUS GROUP OF EQUALS. THERE IS NO LEADER OR DOMINANT ANIMAL, NO PECKING ORDER BASED ON AGE OR SIZE.

The members of a pride pay close attention to one another and work together as a team. Here several lions share a buffalo they have killed.

lioness takes risks and uses energy to hunt large prey but then shares the meal with other lions that may not have helped in the hunt. She is somewhat like a worker bee that forages all day to bring food to the hive and helps raise other bees but dies without ever reproducing herself. For the lion—and the bee and the many other animals that do their part for the greater good—a subtle benefit is to be had in communal life. There is a trade-off between helping sisters and going it alone.

On balance, there is probably a greater advantage for a lion to live in the security of a group and divide kills with freeloading relatives than to live alone and risk yielding kills to strange lions or other large predators. The same balance explains why brother lions help each other drive off rival males but do not compete with each other for females. Since the males in a pride share genes and the females in a pride share genes, why worry which of them actually pair up on any particular occasion? The reciprocal benefits of pride life help related lions raise more young, so the genes that predispose this species to cooperation are perpetuated.

The pros and cons of cooperation are especially clear for male lions. Whereas groups of females do most of the hunting and all of the cub rearing, the adult males in a pride concentrate their energies on defending their breeding rights. Their main job is to exclude strange males from siring cubs with pride females. The larger the group of males, the longer they remain with the pride. The optimal number of males in a pride is three or four. Males in larger groups are almost always related, so kin selection is at work again. If a male that doesn't successfully breed helps defend his pride against competition, he can ensure that some portion of his genes will be passed on to the next generation via his close male relatives.

FACING PAGE

A small cub becomes passive when carried by a lioness and hangs loosely from her jaws. Cubs cannot walk well until they are about three weeks old.

A fallen tree trunk makes a good lookout for a group of lionesses. Two spot something in the distance, while the third takes a nap.

Chapter Two
THE LION'S WORLD

FROM BIRTH TO DEATH

The act of mating is an exhausting one for lions—a pair may copulate three hundred times over a few days. When a female comes into estrus, the male of the moment stays close to her heels, waiting for signs of cooperation. Courtship may start with mutual rubbing of heads and flanks, accompanied by low rumbling sounds. When she is ready, the lioness crouches in front of her mate and flicks her tail up. Like their small domestic relatives, the big cats indulge in rough and noisy pairing. He typically grabs the back of her neck in his jaws when mounting, which may both deter a change of mind and help produce the relaxation reflex seen in kittens held by the scruff. A few seconds later he ejaculates with a loud yowl, then quickly removes himself or risks being turned on with a snarl and a swat of a paw. For her part, the female concludes by rolling over.

After taking only ten to fifteen minutes to catch their breath, the couple is ready for a repeat performance. Their amorous escapades continue day and night through the estrous period, which lasts from two to five days. Despite this Olympian effort, most females give birth after only one of every four or five estrous periods. In other words, a lioness may copulate well over a thousand times before producing one litter of cubs. Nonconceiving estrous cycles are most common just after a pride has been taken over by new males and may represent a trial period of testing the newcomers' virility and stamina. Because females outnumber males in a pride by two or three to one, a breeding male may spend about one-fifth of his time while in a pride in pursuit of

FACING PAGE
Lion prides vary in size from four or five animals to twenty or more.

sex. A final statistic on this subject—a successful adult male will copulate about twenty thousand times during his lifetime.

There is no breeding season for lions. Like all cats, female lions are induced ovulators; they need the stimulus of repeated copulation to release eggs from their ovaries. With no regular estrous cycle, a lioness may come into heat in as little as two weeks or as long as several months after her last estrus. Once pregnant, however, a lioness suspends further mating until she has raised her cubs to about two years of age. The death of her cubs will trigger a new cycle.

Within a pride, all or most of the lionesses of breeding age tend come into heat around the same time. Such synchronized estrous cycles are common in many social species, including humans. Synchrony is most commonly induced by the takeover of a pride, when incoming males kill all the young cubs, causing the mothers to come into heat.

Pregnancy is short (about 110 days), and a mother-to-be does not gain much bulk. She gives birth in a sheltered den among rocky crevices or in thick brush, producing anywhere from one to four cubs. Helpless as kittens and blind at birth, lion cubs are nurtured and protected by their mother in the den for five to seven weeks, until they are robust enough to walk and join the rest of the pride. In contrast to the helplessness of these newborn carnivores, baby calves and foals are able to stand and run with the herd within hours of their birth.

When several females give birth within a short time of each other, their cubs are reared together and may suckle from any lactating female in the pride. The cubs can also be left in the care of one or two adults in a creche while the others go off to hunt. Synchrony pays off, as litters born together tend to have a better food supply and a higher rate of survival.

FACING PAGE
A breeding pair of lions copulates many times over two to five days.

Communal nursing is not unique to lions. Other social carnivores, as well as some species of rodents and pigs that live in small kin groups, also share maternal duties. Why nurse another animal's offspring? The amount of milk a mother produces depends more on her food intake than on the number of cubs she has. Mothers of small litters produce more milk than they need for their own young and can afford to be generous to another mother's cubs. If the other mother is a close relative, the lioness who helps with nursing also helps promote her own genetic line. Cubs are nursed until they are six to eight months old and then depend on the hunting success of their pride until they have learned to kill for themselves, around two years of age.

The role of chance looms large in the lives of every living thing, and it is never larger than at life's debut. Most of the wide-eyed cubs that tumble from their birthplace into the bigger world are not destined to have offspring of their own. Food is the key to survival, and the pride is the key to food.

By the age of four months, cubs can follow adults to nearby kills, where they begin licking at blood and sampling small scraps of flesh. Cubs born later than others in their pride have a serious disadvantage. The younger animals lag behind when the pride is on the move and are shoved aside from kills or robbed of food by the older cubs. Many die of starvation around this time. Other cubs may be injured in brawls at the carcass or become separated from the pride and picked off by hyenas or leopards. Depending on conditions—especially on the supply of food—a pride may rear most of its cubs or lose most of them. On average, between two-thirds and three-quarters of cubs do not make it to adulthood.

In the face of their uncertain food supply, young lions are resilient. Days of hunger can reduce them to listless bags of skin and bone, but a big buffalo

FACING PAGE
Several females in a pride may give birth within a few days of each other, and cubs reared together can suckle from any lactating female in the group.

carcass quickly restores them. Like plants given water, cubs are filled out by a good meal, and second helpings make them plump and playful once more. Given plenty of food, cubs grow fast, but it is hard to estimate their age from their size. A well-fed juvenile of eighteen months may be as big as an ill-nourished two-year-old.

At two years, youngsters are nearly fully grown. The risks of infancy are behind them, and they are ready for adult business. If there are several animals of this age in a pride, they tend to stick together; if there is only one or two, they begin to integrate with the adults. As cubs, they honed their reflexes through play, grappling and chasing one another in practice for the serious skills of hunting. Now they begin to learn the business of being a predator through trial and error. Young lions are apt to pursue any target that excites their enthusiasm, but in time they come to understand the importance of selection and stalking. Eventually, they graduate to lessons in how to bring down prey and kill it.

Most young females will live out their lives in the pride in which they were born, joining their relatives on the hunting team and beginning to breed by the time they are about four years old. Their first litters tend to be small and have a low survival rate. If the pride is especially large, some lionesses may disperse at two to three years of age and form new prides of their own.

Young males start to fill out their manes at the age of three. At thirty months, they are already larger than adult females but not yet as big as the mature males. They are driven from the pride where they were born when new adult males take over their fathers' tenure. Brothers born together may remain together throughout their lives. They live as freely roaming nomads

FACING PAGE
Young cubs follow the pride to nearby kills. They slowly begin to add meat to their diet of mother's milk from two months of age and are fully weaned by about seven months.

Young male lions are driven from the prides in which they were born and spend a few years wandering as nomads before taking over a pride of their own when they are about four or five years old.

for a time after saying good-bye to the pride, occasionally teaming up with other independent males for periods during their bachelor years.

Males generally establish their breeding tenure in a pride when they are about four or five years old. This may come about the easy way if they find a pride with no resident males—perhaps because the males have died or have left to find better hunting grounds. More commonly, they must challenge and displace the current males, using their superior strength or numbers.

The takeover of a pride by new males can be a protracted affair, full of maneuvering. The incoming male team may patrol the edges of the pride territory for weeks, playing a battle of nerves, filling the nights with their roars. They gradually extend their influence over the area until their persistence pays off. The final outcome depends on the relative strength of their adversaries. One day, seeing the writing on the wall, the incumbents may leave, although perhaps not until after a token skirmish. If the opponents are evenly matched, there are savage fights, which can result in the deaths of some males. The successful young males now enter the key breeding period of their lives.

Although male coalitions are a key to breeding success, there is an alternative breeding strategy for solitary males or males whose coalition partners have been killed. They often become "satellite males," occupying border areas between pride territories and having brief affairs with females from local prides. As long as these males are not caught by the resident males, any cubs they father will be accepted by the pride as its own.

Shortly after a change in tenure in the pride, the new males commonly kill any cubs under eighteen months old and drive out all young males. In this way, they ensure their paternity during their short reproductive lifetime, which is generally only as long as their tenure in the pride. This is their only opportunity to pass on their genes, and they cannot afford to put any effort into helping raise other males' offspring.

Adult males that are expelled from a pride have little to look forward to. They become nomads again, but their size and age make hunting difficult and most die within a year or two of expulsion. With a life expectancy of about eleven years, males have a shorter life span than females.

When a lioness hits her fourteenth year, her reproductive days are over. Her teeth are blunt, broken, and yellowed. Her ears are tattered, her snout battered, her eyes sunken. Even though she may not produce more cubs or help with the hunting, however, she still has an important role as chief babysitter. A wild lioness may live as long eighteen years before succumbing to disease, starvation, injury, or hyenas. In captivity, with good food and medical care, lions can live for thirty years.

THE LAND

An abundance of predators is a sure sign of a land's ecological richness. For each predator, there is a wealth of prey; for each plant eater, a cornucopia of vegetation. Every predator sits atop a pyramid of life, shaped and supported by the tightly knit food webs spread out below. Nowhere is the table set more richly than in East Africa, where over forty species of carnivorous mammals, large and small, carve up the land's animal protein between them.

Among the larger hunters, the lion is the biggest predator in a tough environment where gentler beasts run the risk of ending their lives in the jaws of hyenas, leopards, hunting dogs, cheetahs, jackals, or foxes. Each of these predators has its own niche in the bush and grasslands of Africa, and

AN ABUNDANCE OF PREDATORS IS A SURE SIGN OF A LAND'S ECOLOGICAL RICHNESS. FOR EACH PREDATOR, THERE IS A WEALTH OF PREY.

only the hyena comes close to challenging the lion's place at the head of the carnivore club.

The lion's preeminence is due to the combination of its size and sociability. At thirty times the weight of an average tabby, one lion is daunting enough. A tag team of a half-dozen or more of the heavyweight hunters puts discretion ahead of valor for most creatures. A full-grown elephant is one of the few animals that can face down a pride of lions.

Their diet is bloody, but all flesh is grass; so the lion's story properly begins with the land and vegetation. For choice of home, lions prefer places where there is cover for hunting and denning. They can be found wherever there is enough prey—from open woodland or thick bush to savanna. They range up to 3600 meters (12,000 feet) above sea level on the slopes of mountains and survive even in the dry conditions of the Kalahari desert.

The classic East African environment is seen in the Serengeti, 13,000 square kilometers (5000 square miles) of rolling hills. Two-thirds of the Serengeti is open woodland (with few or no shrubs or underbrush), and one-third is grassland. Acacias (thorn trees) up to 9 meters (30 feet) high dot the savanna, and fig, wild date, and sausage trees mark watercourses. Rains come seasonally and erratically, and only the larger streams provide water year-round.

The soil of the East African plains is fine and porous, formed of volcanic dust and ashes in thick layers that have formed over the ages along the flanks of the Great Rift Valley. This 80-kilometer (50-mile)-wide gash marks the place where the eastern edge of Africa is being torn away from the rest of the continent by the movement of tectonic plates. Kopjes, or granite outcroppings, rise from the plains. They are the bald heads of old hills that have

FACING PAGE
Lions are found in a variety of habitats, from open savanna and semidesert to woodland or thick bush. Their range depends mainly on hunting opportunities in the area.

outlasted the softer material that once surrounded them. Now, in their turn, they are slowly yielding to rain, wind, and gravity.

The raw materials of life are light, air, and water, transformed by the alchemy of photosynthesis into the living matter of plants. The crucial part of this equation in the semiarid regions of East Africa is water, sucked from the Indian Ocean by the hot sun and hauled to the plains in bloated black clouds chivvied westward by the prevailing winds. Deliveries of rain are strictly seasonal, and the spectacle of life on the savanna is finely balanced according to the timing and extent of the rains.

A flash and rumble on the horizon heralds a time of feasting for all. Heavy skies dump their cargo onto the dust, and in stygian sanctuaries below the sun-parched surface, seeds and roots revive with the touch of moisture. Heavy raindrops sink quickly into the mineral-rich soil, to be captured at once by plants long adapted to a regime of dryness and deluge. Armed with water, new green shoots make their debut into the world of light and air. The race for survival is on, with competition among the plants as keen as any among animals.

A FLASH AND RUMBLE ON THE HORIZON HERALDS A TIME OF FEASTING FOR ALL. HEAVY SKIES DUMP THEIR CARGO ONTO THE DUST, AND IN STYGIAN SANCTUARIES BELOW THE SUN-PARCHED SURFACE, SEEDS AND ROOTS REVIVE WITH THE TOUCH OF MOISTURE.

Plants use different strategies in the rush to reproduce and claim territory for their descendants. Some stay low and spread out runners to cover as much open ground as possible. Others spring up thick and tall to shade and choke their neighbors. Grasses surrender their seeds to passing winds, which carry the new generation to distant patches of bare ground. Spreading roots extend their grip on loose sand or specialize in drawing water from the soda-saturated soil around a shrinking waterhole.

The wet season runs from December to May, nurturing a sea of meter-high (3-foot-high) grasses topped by a spume of spiky seed heads. When the dry

FACING PAGE
Rain clouds signal a time of feast for hunters and hunted alike.

season returns, plant life shrinks once more into roots and seeds. Grass fires race through skeletal stems, leaving piles of ashes in their wake. Dung beetles disperse the undigested remains of a million meals, while termites work decaying matter into the soil. All is prepared for the next rains.

The flora of the savanna must not only compete with one another to survive but also defend themselves from plant eaters. Their arsenals are as diverse as the species. This one is unpalatable, with a taste like turpentine; that one has tough and indigestible fibers; and a third stores abrasive crystals in its tissues to give it the texture of sandpaper. Using its enemies to advantage, a pragmatic plant called mat-grass does best when grazed. Like a lawn, the low-growing grass springs up from its base, not from the tips of its shoots. Grazers help it by eliminating nearby competitors and clearing space, letting the plant spread to form a soft, springy rug. In the absence of grazers, mat-grass is overshadowed by other species and is unsuccessful.

Like the plants it feeds on, each herbivore on the grassland has its niche, and the variety of survival strategies among plants is mirrored by the variety among the animals that eat the plants. There are short-grass feeders and long-grass feeders, as well as resident populations and migrants that move with the seasons. How vastly different from this complex and efficient ecosystem is the crude and simple system of farming with which people seek to replace it, substituting a handful of alien crops and livestock for the hundreds of native species that the land naturally supports.

Wildebeests are one of the most numerous grazers on the savanna. They are lawn mowers, pressing sheets of grass blades together between their wide, flat lips and then cutting them neatly with a row of sharp lower teeth. Small and dainty Thomson's gazelles, another common grazer, enforce conformity,

FACING PAGE
Impala keep a sharp lookout for lions. When pursued, they flee with spectacular twisting leaps and bounds.
BIOS (DENIS-HUOT)/ PETER ARNOLD, INC.

nipping off any shoots that stick up above their fellows. Coarser grasses left by these two are consumed by zebras, with their broad, horsey teeth. Lacking the ruminant system of the cud-chewing antelopes, zebras must ingest a greater volume of food to compensate for their less efficient digestion. They stuff themselves, looking fat and round compared with the slim and elegant antelopes. Another grazer on the savanna is the eland. Largest of the antelopes, eland are always on the move, seeking out patches of soft young grasses and herbs that are scattered in patches. Although it takes work to find these plants, the eland's meals are more nutritious than regular grasses.

In the long dry spells between rains, the grasses of the Serengeti savanna shrivel and the vast grazing herds migrate to the open woodlands, where they join other grazers, such as buffalo and impala, as well as browsing giraffes, rhinos, and elephants. With the migration of the wildebeests (about one million), Thomson's gazelles (400,000), zebra (300,000), and eland (7000), the population of grazers on the Serengeti shrinks by about 80 percent. Other herbivores of the plains, such as topi, kongoni (Coke's hartebeest), Grant's gazelles, and warthogs, are year-round residents, eating anything they can find, including old, tough grasses and infrequent shoots and herbs. After especially long dry spells, even these grazers must move to greener pastures.

All these animals are food for carnivores. As populations of prey appear and disappear, disperse and concentrate, according to the weather and time of year, the hunters must adapt their eating habits and behavior to the shifting scene.

In the dry season, according to estimates made by Schaller, the weight of grazer flesh in the Serengeti lion's domain falls by nearly two-thirds. Prides residing in the grassland, which maintain their hunting range year-round,

FACING PAGE
As potential meals for several species of carnivores, grazing animals such as zebra find safety in numbers.
DOUG CHEESEMAN/PETER ARNOLD, INC.

subsist on gazelles, ostriches, and warthogs, while their woodland cousins reap the bonanza of the incoming herds. In the wet season, their fortunes are reversed as the herbivores return to the plains. There are also nomadic lions who follow the herds from one habitat to the other but pay the price of hostility from the prides whose territories they cross. These nomads are old males, young males just starting out on their own, and females not belonging to a resident pride.

During much of the year, the predator's life is hard work. Each meal must be found, stalked, and brought down. Each meal does all it can to resist being found, stalked, and brought down. Imagine, then, how much harder the hunt is for a predator who is pregnant, and how critical it is for one who is suckling growing cubs. What better time to give birth and raise cubs than when the lions' prey are having *their* young, when there are heavy mothers and helpless infants to provide easy meat for the family?

For the herds of grazing animals, safety lies in numbers, vigilance, and speed. With so many predators around it is fatal to stand out from the crowd, and pregnancy is synchronized within the herd to improve the odds of survival. The peak of births usually comes with the rains, when the plains are newly filled with fresh food and lanky calves can be hidden among the high green grasses. Thousands of calves first see light within a week or two of each other, and those born very early or very late are at a disadvantage.

FOR THE HERDS OF GRAZING ANIMALS, SAFETY LIES IN NUMBERS, VIGILANCE, AND SPEED. WITH SO MANY PREDATORS AROUND IT IS FATAL TO STAND OUT FROM THE CROWD.

THE HUNT

Lions, for the most part, spend no more than three or four hours each day on their feet, passing the remaining twenty hours lying down with their eyes closed or gazing vacantly into the distance. It's not that lions are lazy, but

FACING PAGE
Adult males in a pride leave most of the hunting to females. The males' main role is to protect their territory from other males.
GAVRIEL JECAN/ ART WOLFE INC.

they must conserve their energy in lives adapted to the possibility of feast or famine. Like many carnivores, lions can go without eating for several days and then fill their bellies with a gargantuan meal. Downing 30 kilograms (60 to 70 pounds) or more of meat at one sitting, a well-fed lion gains more girth than a pregnant one.

A lion on the African plains appears to have its pick of meat on the hoof, but obtaining a meal from the passing parade isn't as straightforward as picking berries from a bush. Most of the animals that the lion sees from its hiding place are safe from its jaws. Many of them, like the dik-dik and hare, are too small to bother with. Others, like the hippopotamus and elephant, are too large. Some have dangerous defenses, like the porcupine's long quills and the baboon's daggerlike canine teeth. Many are defended by their speed, and a few, like the warthog, take shelter in burrows. Like most hunters, the lion pursues what is most rewarding for the least effort and generally avoids anything that might put its own well-being at risk.

The lioness breaks from cover as though loosed by a massive elastic band stretched between her and the galloping zebra, eliminating the distance between them in three heart-pounding seconds. Her startled prey, an unwilling recruit in this ancient drama, kicks behind him in desperation, then swivels sharply away with a screaming bray into a collision course with a second lioness closing in from his right. It is, as kills go, fairly quick. One of the 150-kilogram (330-pound) cats hangs by her jaws from the zebra's throat, stopping his breath as effectively as a knot in a balloon prevents air from escaping. His nostrils flare in vain, his eyes roll, and strings of blood-tinged saliva dribble from the side of his mouth onto his killer's fur. In shock, he probably doesn't feel the second lioness rip a hole in his belly or a third

FACING PAGE
Hunting lions use ambush or slow stalking to get close to their prey before making the final rush.
STEPHEN J. KRASEMANN/PETER ARNOLD INC.

pounce on his back and flay his tough hide with her claws. Her weight finally topples him onto his side, where he stirs the dust with reflexive kicks as the predators begin their meal.

The pandemonium of hooves and honks that shook the plains only minutes before is replaced by muffled snarls. Nearby grazers stop briefly to look at the carnage, then rejoin their moving herds. Death is a daily game of chance for these plant feeders, but the cards are not dealt evenly; losers are marked by signs that let the hunters sort them from the pack.

The long association between predators and prey has the intimacy of a marriage. A lion must study the habits of the creatures it eats with the assiduity of a lover if it is to stay well fed. No point in wasting time and energy on a grazer with a quick eye, sharp ear, and fleet foot. That one with the ribs showing is a good candidate for dinner, and the lost youngster who calls for her mother is almost certain not to pass on her genes to the next generation. Potential prey, too, must know their enemies. A lamb may lie down near a lion with impunity if that lion is bloated with food and wanting a snooze. It is not uncommon to see gazelles strolling within plain view of a lion pride, albeit keeping a wary eye on the big cats. Lions are safer when they can be seen.

The keys to the hunting behavior of lions are cooperation and flexibility. Larger prides take advantage of their numbers to organize ambushes. In some cases, individual lions may specialize in different tasks in the hunt, according to their different skills. For example, lighter, swifter animals may take on the chase, while their larger, heavier sisters wait in hiding to bring down the prey. The same lioness may also switch roles according to the prey type, taking a lead in warthog hunts but playing follower when the game is buffalo.

FACING PAGE
A lioness clamps her jaws tightly onto the throat of a topi to suffocate it.
BIOS (DENIS-HUOT) / PETER ARNOLD, INC.

A newspaper cartoon shows a king on skis (wearing his robes and crown) grumpily addressing a ski instructor. "But I'm a King!" he protests. "Surely I don't have to start out with baby snowplow turns." The king of beasts may be a top predator, but it, too, must learn from scratch. Because young lions find out all they know about hunting from their pride, there may be "cultural" differences in prey selection and killing techniques passed on in different areas with different conditions. For example, one pride of lions living along Namibia's desert coast learned how to prey on Cape fur seals. Another pride, in the flat, open terrain of northern Namibia, specializes in catching springbok, one of the fastest of all antelopes. These lions' hunting strategy is as highly coordinated as a military exercise, with flanks and centers moving together to encircle their prey. This teamwork is unlike the generally more haphazard hunting style of lions in the open woodland habitat of the Serengeti.

For the best return on energy invested, a pride prefers larger prey. A group of lionesses will stalk through a herd of gazelles to catch a zebra or wildebeest, combining efforts to bring down quarry weighing up to four times as much as a lion. Single lions catch prey ranging from half their weight to twice their weight. Scavenging or driving smaller predators away from their kills requires the least effort, and in some areas lions get a third or more of their food in this way.

Lions are opportunists, and they have been known to eat such oddities as crocodile, guinea fowl, leopard, aardvark, monkey, python, catfish, elephant, locust, and tortoise. But although chance and season influence the content of a lion's meal, the bulk of their diet usually comes from a core of no more than five species. Their most common prey are zebra, wildebeest, roan,

FACING PAGE
Lions eat so rapidly that a large pride can reduce the body of a zebra to a skeleton in half an hour.
ERWIN & PEGGY BAUER

ABUNDANT PREY MAKE HARMONIOUS PRIDES, WHERE ALL TAKE THEIR FILL AT EACH KILL. BUT SCARCITY BRINGS COMPETITION, IN WHICH EVERY LION MUST FEND FOR ITSELF.

sable, springbok, gemsbok, kob, impala, Thomson's gazelle, topi, warthog, buffalo, waterbuck, and kongoni (Coke's hartebeest).

Abundant prey make harmonious prides, where all take their fill at each kill. But scarcity brings competition, in which every lion must fend for itself. With size and strength weighing in the balance, males get the better of females, and youngsters get short shrift from their elders. Cubs are always at a disadvantage in the snarling scrum around a carcass and are the first to suffer from starvation if prey are scarce.

In theory, if not always in practice, human cultures give priority to their young. In nature, the young are the most dispensable. There would be no benefit to a pride in letting cubs feed first, for the survival of the group depends on the hunting skills of adults. A weak hunter is of no use to either herself or her offspring.

FACING PAGE
Smaller prey, like this gazelle, may be dragged away to shelter, where it can be devoured in peace.
BIOS (DENIS-HUOT)/ PETER ARNOLD, INC.

Lions spend about 80 percent of their time resting.

GAVRIEL JECAN/ ART WOLFE INC.

Chapter Three

LIONS AND PEOPLE

Cave walls in France and Spain bear the likenesses of lions scratched into the rock by our forebears more than fifteen thousand years ago. At that time, lions ranged from southern Europe to India and were widespread over much of Africa and Arabia. They maintained most of this realm until just before the beginning of the Christian era, when rising kingdoms around the Mediterranean and Far East came into increasing conflict with the king of beasts.

Words cut in stone in the ancient kingdom of Assyria (now northern Iraq) during the reign of King Ashurbanipal in the 7th century B.C. vividly record an unhappy relationship: "The hills echo with the thunder of their roars . . . The herdsmen and their masters are in distress. Women and children mourn . . . On my hunt I have entered the lions' hiding places. I have destroyed their lairs."

Using mastifflike dogs, the Assyrians hunted lions around their settlements at Nineveh, on the east bank of the River Tigris, and left stories and pictures with a freshness of detail. The inscription above is part of a longer tale from a relief now housed in the British Museum. It describes how heavy rains produced impenetrable thickets in which lions took refuge and how, as their numbers grew, the lions emerged at night to kill both livestock and people.

This writing on the wall sets the tone for the next two millennia. Lions could not coexist with expanding civilizations. In nearby Egypt, hieroglyphics record that Pharaoh Amenhotep III (1417–1379 B.C.) killed

FACING PAGE

The range of the lion in Africa today has been confined to smaller and smaller areas by a spreading human population.

102 lions in the first ten years of his rule, hunting with bow and arrows from a chariot. Like the Assyrian records, the commemorative scarabs that immortalize this feat have a boastful quality. It was a prerogative of kings to hunt a powerful hunter, and owning a lion-skin rug clearly transmitted high status to its owner then as now.

Lions returned the favor and attacked the baggage camels of the Persian ruler Xerxes I during his venture against Greece in 480 B.C. Writing of this encounter, the great historian and traveler Herodotus thought lions were common in the region at that time. Nearly two hundred years later, Aristotle thought them rare. Aristotle was tutor to Alexander the Great, another ruler whose name is linked to lions. Alexander's portrait on coins is stylized as the head of Heracles wearing the skin of the Nemean lion, the fearsome specimen killed by the legendary hero as the first of his twelve labors.

The main entertainment of the Roman circuses was chariot racing rather than public execution by tooth and claw, but the memorable matching of lions and Christians at the Circus Calligulas under the Emperor Nero in the 1st century A.D. still echoes loudly in our historic memory. Roman bloodletting, both animal and human, was carried out mainly in the Colosseum, where rowdy crowds of up to 50,000 paying spectators assembled to enjoy the slaughter. In one celebration, 2000 gladiators and 230 wild animals were billed to die. It is not known if that ratio was due more to the relative value of human and animal life at the time or to their relative scarcity.

Lions had almost been eliminated from Europe by A.D. 100 and were transported, together with other exotic beasts, from North Africa for the Roman festivities. The "hunting" of wild animals in the Colosseum continued until A.D. 523. Lions survived in Palestine until the Crusades of the

FACING PAGE
The Repast of the Lion *by Henri Rousseau.*
THE METROPOLITAN MUSEUM OF ART, NEW YORK

12th century and in Turkey until the late 19th century. They were still occasionally encountered in their last outposts in Syria, Iraq, and Iran in the 20th century.

THE MYTHOLOGICAL AND SYMBOLIC LION

Both feared and admired, the lion looms large as a leading character in the stories of many cultures. Leonine stardom was literally established long ago by the naming of the constellation Leo. The Greeks connected this combination of stars with the Nemean lion slain by Heracles. Because the lion is a symbol of heat and passion, its constellation occupies a part of the zodiac through which the sun passes in late July and August.

Why is a lion the most popular poster animal for depicting strength, courage, vigor, and regal status? Why do we "lionize" this large cat and grant him (rarely her) the lion's share of exposure on heraldic insignia? Why are lions preeminent among the animals that lend their names to popes, emperors, saints, clubs, and soccer teams? It cannot be merely their large size or predatory habits that impress us, for other species are large and predatory. It is the combination of trappings and demeanor we admire: the flowing mane, the commanding roar, the languid aura of confidence and superiority. Other animals are more dangerous, but none have quite that look: the "lazy, lordly power born of the carelessness of authority."

FACING PAGE *Resting from the heat with her mouth open, a lioness protects an unfinished meal from scavengers.* GAVRIEL JECAN/ ART WOLFE INC.

The lion projects its image even when far from its native home. In London's Trafalgar Square and outside the New York Public Library, statuesque lions stand sentinel to the glories of empire and the power of knowledge in countries where they never lived. The Chinese gave the lion a prominent place in traditional mythology and festivities. Actual lions made

ABOVE

Two Lions Licking Cubs, *from* The Workshop Bestiary. *In medieval myth, lion cubs were born dead and brought to life after three days by being licked by their fathers.* THE PIERPONT MORGAN LIBRARY/ART RESOURCE, NEW YORK

their way to China via the famous Silk Road, which linked the Chinese and Roman Empires before the start of the Christian era. To obtain rights to trade for silk along this route, rulers in what is today Iran and Afghanistan sent exotic gifts, including lions, to the Chinese emperors.

The lion dance seen in Chinese communities around the world today dates back two thousand years to the time of the Han Dynasty (206 B.C. to A.D. 220). The dance is connected with the tale of a strange and terrible mythical creature that ate humans and animals. Called *nien* (which sounds like the Chinese word for "year"), the creature could not be defeated by either fox or tiger, which are indigenous to China. In despair, the people asked the lion for help. The lion shook his mane, rushed at the *nien,* and wounded it. Running away in humiliation, the *nien* declared it would return for revenge.

FACING PAGE

The road to adulthood is risky for lions, and a high percentage of cubs fall victim to starvation, disease, or accidents, or to the jaws of hyenas or strange male lions.

A year later, the *nien* did return, but the lion was now too busy guarding the emperor's gate and could not help the people. The villagers decided to do the job themselves. Two men, skilled in martial arts and dance, disguised themselves as the lion inside a cloth and bamboo costume. They pranced, shook, and feinted an attack on the monster, causing it to flee once more. The lion still dances on the eve of every Chinese New Year and other important occasions, frightening away evil for yet another year. This dance is sometimes confused with the dragon dance, but the dragon is performed by more than two people.

The lion's ferocity is a legendary quality with the greatest staying power. For example, the story of the lion's share, recorded in the 6th century B.C. by Aesop (possibly himself a legendary figure), describes a rather intimidating lion who takes much more than he is due after joining in a hunt with several other kinds of animals.

In contrast to this bully, however, is the grateful lion encountered by Androcles. The famous tale is told by Aulus Gellius, a Roman of the 2nd century A.D. Androcles was an escaped slave who earned the lion's gratitude by taking a large thorn from its swollen paw. Shortly after, Androcles was recaptured and was to be killed by wild beasts in the circus. One of the beasts, however, turned out to be his old friend. When the lion ran up to Androcles and began licking and nuzzling the slave, the amazed authorities freed them both. Once more the lion is a symbol of strength, but the moral lesson here is that those with power also have obligations.

With power to spare, even a part of a lion is a potent symbol. A lion's body joined to a human head makes a sphinx. Its paws added to a dragon result in a manticore. Its hind legs and tail grafted to the head and wings of an eagle create a griffin.

FACING PAGE
The premier animal symbol of power and strength begins life with a cute face.

AROUND THE MIDDLE OF THE CENTURY, A NEW BREED OF ADVENTURER ARRIVED ON THE CONTINENT FROM THE PUBLIC SCHOOLS OF ENGLAND. FOR THESE FELLOWS, HUNTING WAS NOT MAINLY A MATTER OF KILLING BUT MORE OF A SCIENTIFIC AND INTELLECTUAL PURSUIT.

The opposite of the fierce lion may be the cowardly lion. Among African tribes that traditionally hunt lions, stories are told of the fainthearted predator slinking from a warrior armed only with a spear. And the blustering beast that Dorothy encounters on her journey to the Wizard of Oz is a lily-livered lion who eventually gains his valor with a medal.

PESTS AND TROPHIES

The dominant Western view of lions from the mid-1800s to the mid-1900s was the view along the barrel of a rifle. Much of the favorite reading material about lions and other African fauna during the early part of this period was produced by explorers and hunters, neither of which regarded predators with a friendly or academic eye.

In the pioneering years of the European colonization in Africa, settlers and nomadic Masai alike saw lions as threats to their livestock. Big cats with a taste for beef or goat meat did not long survive the spears of the Masai or the bullets of the Europeans. Shot as pests, lions were cleared from large areas of southern Africa during the Boer expansion of the early 1800s.

More serious than potshots aimed at feline poachers, however, was the wholesale clearing of the land and its native grazers to make way for farms. Thanks to greed, lack of laws, and the .450 Martini breech-loading rifle, a wealth of wildlife was slaughtered in South Africa in only a few decades. By the time the English explorer Frederick Selous journeyed there in the 1870s, elephants and rhinos were already scarce or extinct in many areas, and the enormous herds of zebra, gazelles, and other herbivores that had been present fifty years earlier had dwindled to scattered groups. With their prey gone, the carnivores vanished soon after. "The high veldt of the Transvaal . . . is a dreary waste for the naturalist or sportsman," wrote Selous.

Around the middle of the century, a new breed of adventurer arrived on the continent from the public schools of England. For these fellows, hunting was not mainly a matter of killing but more of a scientific and intellectual pursuit. Sportsmen, collectors, and gentlemen, they hunted by rules, with an eye to fair play. Africa was still teeming with wildlife, but the gentlemen adventurers placed rigorous limits on their sport. Honor demanded that they dismount from horseback to face their quarry on foot, and in truth their undertaking demanded nerve, skill, and some luck. A misfire with the weapons of the period left them little margin of safety from a wounded and dangerous beast only paces away, and many of these hunters fell victim to their intended victims.

This style of plucky hunter could still be found well into the 20th century, but his heyday did not last long. By the 1880s, a visit to Africa was no longer the preserve of the very rich and intrepid. Somaliland (now Somalia) on the Horn of Africa was cheap and accessible by boat, with lions waiting to be shot just a camel ride away from the Gulf of Aden. The status symbol of a dead lion was now within the scope of thousands, and Somaliland came into vogue. Army officers, government officials, aristocrats, and businessmen were the mainstay of the first safari floods through the 1890s, and they popularized big game hunting in Africa. Within a few years, the country was largely shot out.

The reading public developed a taste for safari stories. Theodore Roosevelt's *African Game Trails* was so popular that it was reprinted twice within three months of its first publication in August 1910. A preface plate shows the hunter, one hand on his hip, the other on his rifle, standing just behind the body of a large and spectacularly maned lion. In the romantic and guileless language of his time, Roosevelt describes his experiences with creatures of foul and evil ferocity in a land of dread brutes and swarming foes, his encounters with savages and hideous spotted beasts. Above all, however, he finds a land that "teems with beasts of the chase, infinite in number and incredible in variety."

Such images fueled the Western imagination during the late Victorian era and lingered for decades after. Big game hunters have long since fallen out of favor, recast by our age as anachronistic, mustachioed villains. The early 1900s brought ivory hunters and big money, automatic weapons and motor vehicles. Rules soon flew out of the window, and fair play was out of fashion. On the flat, high plains around Nairobi, lions were ridden down on horseback. They were pursued with packs of hunting dogs. They were drawn to bait and shot from enclosures. They were massacred from vehicles, which they didn't recognize as dangerous. Superb specimens were converted to floor coverings or wall hangings, and a single safari party might take a hundred lions or more.

Not everybody treated Africa as a shooting gallery. Many appreciated the spectacle of wildlife and were concerned about conservation even in the 1800s. Officials recognized the folly of a perpetual open season with no limits, and the first system to license hunting was introduced in British East Africa in 1905. In 1937, licenses issued by the Kenyan government restricted each hunter to four lions, a limit later reduced to two.

World War II moved worries about wildlife to the back burner. Britain had to feed a quarter of a million Italian prisoners of war that it held in

FACING PAGE
By ten years of age, a male lion is past his prime and will probably live only another year or two.

Africa, and supplies of beef and mutton were short. The government turned to game meat, and in 1944, seventeen thousand zebras and unknown numbers of impala, eland, and gazelles were turned into prison food. Ironically, the initial effect was to increase the lion population. An abundance of carrion improved the survival rate of cubs and young lions, usually the first to starve when food is scarce. The predator population started to climb just as the herbivore numbers fell, with predictable consequences. The "plague" of lions turned their attention to newly established herds of beef and dairy cows and sheep, and a major task of game control officers in East Africa during the 1940s was to protect livestock from lions.

MAN-EATERS

Unarmed modern humans must be the easiest of prey: thin-skinned, slow, weak, and without horns, hooves, or tusks. Anthropophagus (man-eating) lions in Africa have, in the inimitable words of Theodore Roosevelt, "always offered the chief source of unpleasant excitement." Colonel J. H. Patterson's classic book, *The Man-eaters of Tsavo,* sent chills through generations of readers with its descriptions of a nine-month reign of terror in Kenya in 1898, when twenty-eight Indians and an unknown number of Africans working on the construction of the railroad from Mombasa to Lake Victoria were dragged away in the night and eaten by two lions.

FACING PAGE
Close encounters between lions and people often end fatally for one or the other.

The wonder is that lions don't try to eat people more often. In fact, unarmed researchers on foot have found that lions usually run away from them if they approach in plain view. The biggest danger comes from stumbling onto resting lions hidden from sight. In one of his books, pioneer lion researcher George Schaller describes an episode in which he was following a trail that at one place descended into a deep ravine: "Filled with joie de vivre, I dashed down one side and up the other and almost into a group of nine or more lions. Without stopping, I raced back the way I had come, going considerably faster." He halted on the opposite bank to see that the lions "still sat there, somewhat puzzled."

The lions were probably well fed. In other instances, being in the wrong place at the wrong time can be fatal. Sometimes humans are targeted as food by lions when other prey becomes scarce. When an outbreak of rinderpest wiped out much of the livestock and hooved wildlife in the Ankole region of Uganda in the early 1920s, lions in the area began preying on people and reportedly killed hundreds. Once they have discovered human flesh, individual lions may develop a taste for it. Many recorded examples of man-eater attacks involve repeat offenders that even go so far as to enter villages and drag people from huts.

As a rule, lions are far less aggressive during the day than at night, when they do most of their hunting. Our prehistoric ancestors most likely reduced their chances of deadly encounters with lions by moving about in groups only by day, when lions usually rest, and seeking shelter at night. But still the lion remains a rare reminder of human frailty and one of a select, small group of animals that attack and kill people.

ENTERTAINING AND SOCIALIZING LIONS

The idea of lions as man-eaters was no doubt behind the popularity of lion acts in circuses. Modern circuses came into being in the late 1700s in Europe, but wild animal acts starring lions and elephants were first introduced to the big top in North America. An American lion tamer named Isaac Van

FACING PAGE
Unlike other members of the cat family, lions form close social bonds, which they maintain throughout their lives.

Photographs of performer Clyde Beatty show him enclosed in a cage with as many as forty lions and tigers at one time.

Amburgh was popular in England in the mid-1800s and established the enduring image of a brave man entering a cage of big cats. He is probably the first person to have voluntarily put his head into a lion's mouth for public entertainment. Van Amburgh and the French animal trainer Henri Martin appeared in theaters as well as circuses, mounting theatrical performances with titles such as "The Lions of Mysore" and "The Brute Tamer of Pompeii."

Associated with the circuses were visits "backstage" for an extra fee to see the animals at closer quarters in the menagerie. "We are . . . surrounded . . . by death under its most frightful form; and yet we hold our life as securely as if we were seated by our own hearths," wrote one breathless Victorian visitor to the lion room. The safe thrill was ratcheted up several notches by the ever-popular spectacle of mealtime. Even the young Queen Victoria is reported to have visited several of Van Amburgh's performances and, after one, to have lingered "for the purpose of seeing the animals in their more excited and savage state during the operation of feeding them." The lions were kept without food for three days before her visit so as not to disappoint the queen.

Different styles of lion acts developed on either side of the Atlantic during the early decades of the 20th century. The popular American approach starred trainers brandishing guns, whips, and chairs, apparently in great danger of losing their lives to the savage beasts they held at bay. Photographs of performer Clyde Beatty show him enclosed in a cage with as many as forty lions and tigers at one time. In contrast, European trainers emphasized their skill at transforming foes into friends. Performers would stroke and play with their large cats in the circus ring, allowing just enough snarls and flashes of teeth to impress the audience that not just anyone could perform this feat.

Trading on the ambiguous friendly/fierce persona of the cuddly predator is Leo, the mascot who roars on cue at the opening of each MGM movie. Leo was the prodigy of one of Hollywood's first animal trainers, Volney Phifer. Brought to Tinseltown from the Nubian Desert in Sudan, Leo toured the country with Phifer during the 1920s to promote the movie studio, and he still appears in MGM's logo and trademark. Newspaper accounts record that Leo died in the 1930s, but in the mid-1970s I saw him posing for photographs outside the old MGM Grand Hotel in Las Vegas. Like many other hard-working animal stars, the immortal Leo must have had a stable of doubles, even after he no longer remained to play himself.

As social animals, lions are in many ways more like dogs than typical cats. Lion trainers use these clannish instincts to transfer the big cats' allegiance to themselves. Bonding is easier with a lion separated from its own species at an early age. The most famous illustration of this unlikely relationship is portrayed in the book and the movie *Born Free,* the story of the family formed by the lioness Elsa together with Joy Adamson and her husband, George.

The Adamsons raised Elsa from a cub, but the unexpected twist in their story is not that they were able to socialize the lioness to humans but that when Elsa was three years old, they successfully taught her to live as a wild lion. Having learned the ways of her step-parents, Elsa later made the remarkable switch to the life of her real parents. She survived as a hunter, eventually living with other lions in the wild and raising three cubs of her own. She continued her unique life as an ambassador for two species, bringing her wild-born cubs to meet her human family.

Elsa's story provides dramatic evidence of the importance of learning in the lives of lions. Flexible behavior is a quality of many long-lived social mammals,

including wolves, horses, elephants, dolphins, and people. Humans are the most tractable species of all, making it difficult to determine what, if anything, comprises "human nature."

The Adamsons' love of wild animals was not shared by everyone, and their story did not end happily. Semitame lions were unwelcome neighbors to others in the community, and the Adamsons' efforts to protect wildlife antagonized local game poachers. Joy was killed by a former employee in 1980, and George was shot by poachers in 1989.

THE THREATENED LION

Nobody knows for sure how many wild lions are left in Africa today. Estimates range from 30,000 to 100,000. What is sure is that the numbers and range of lions are vastly reduced from fifty years ago and that the decline continues. The reason is seen in the villages, fields, and pastures that push their way across the land, swallowing more of Africa's wild terrain each year until only small, scattered islands remain as both refuge and prison for the continent's large mammals.

It is only by a chance of history that human encroachment on the lion's habitat did not occur much sooner. For centuries, Muslims from the north and Christians from the coast attempted to carry their cultures into the heartland of Africa but could never establish the populations needed to make this happen. What kept them from settling—the secret ally on the side of wilderness—was a microscopic parasite named *Trypanosoma*.

Trypanosomiasis is better known as sleeping sickness because one of its symptoms is extreme fatigue. Today it still kills hundreds of thousands of people and millions of cattle in Africa each year. Livestock not killed by the

IT IS ONLY BY A CHANCE OF HISTORY THAT HUMAN ENCROACHMENT ON THE LION'S HABITAT DID NOT OCCUR MUCH SOONER.

FACING PAGE
A young lioness practices a side swipe while other pride members dig into their meal.

disease are severely weakened, becoming poor breeders, milk producers, and draft animals to prepare land for crops. Sleeping sickness now affects the inhabitants of one-third of Africa's land area and can accurately be said to have prevented the development of farming and human settlement over millions of square kilometers of sub-Saharan Africa.

The trypanosome parasite is transmitted to humans and other animals by the tsetse fly, a brown, blood-sucking insect found only in Africa. The fly ingests the tiny parasites with its blood meal, and the parasites multiply and develop inside the tsetse's body. When the infected fly takes another meal, it delivers the trypanosomes into the bloodstream of the new host.

The impact of sleeping sickness in Africa provides a classic demonstration of the complex connections among ecosystems and human activities. Trypanosomes are endemic to both dry woodlands and the humid forests bordering lakes and rivers. Their traditional hosts are antelopes and other wild ungulates, which have evolved an equilibrium with the parasites and show no overt signs of disease when infected. Tsetse flies themselves prefer the blood of mammals other than humans, and the spread of the disease to people can be seen as a self-inflicted tragedy—the result of the displacement of wild animals by pastoralists and farmers. Sleeping sickness was unknown in many areas of Africa in the early 19th century, and its initial spread is linked to the inroads of colonialism and the establishment of farms in the forests and woodlands.

The destruction of forests and the elimination of wild animals forced the tsetse–trypanosome duo to turn to cattle and people instead. We are accidental hosts, and the virulence of the disease reflects the fact that our species has not yet had time to evolve an accommodation to the blood parasite, as

FACING PAGE
Nomadic lions wander over the African plains, using areas of land many times larger than the territories occupied by prides.

THE SPREAD OF VILLAGES AND FARMS FRACTURES THE ANCIENT LANDSCAPE, AND WILD CREATURES ARE MAROONED ON ISLANDS OF GRASS AND TREES SURROUNDED BY OCEANS OF HUMANITY.

antelopes have. We have learned too late that flies and microscopic parasites are harder to eradicate than trees and large mammals.

Throughout most of Africa, lions are generally considered serious problems where humans and cattle exist and are therefore becoming increasingly rare outside protected areas. When wild prey is scarce and lions are heavily persecuted, their social structure appears to break down. They are thinly scattered in ones and twos throughout their range and do not form large prides. The smaller group size may be an adaptation to avoid detection by human hunters and allow more efficient stock raiding. Because lions are also scavengers, they are especially vulnerable to poisoned carcasses put out to eliminate predators. Lion hunting is prohibited in a few countries, regulated or restricted in others, and permitted in some. Other countries have no legal protection for lions.

Lion populations that avoid bullets and poison may be at risk from diseases such as rabies and distemper. These diseases affect domestic dogs and spread to wild carnivores where villages encroach on their range. In the Serengeti National Park, for instance, more than one thousand lions (one-third of the entire population) died from canine distemper between 1993 and 1996. Silver-backed jackals, bat-eared foxes, and the rare African wild dogs have also died from the disease. Researchers think distemper is passed on through hyenas and nomadic lions, who are infected by village dogs and then travel long distances and mingle with other predators at kills. The World Society for the Protection of Animals set up a program to stop the disease at its source by vaccinating dogs in the area and provided distemper shots for about fourteen thousand dogs in Serengeti District during 1997.

FACING PAGE
Lions keep watch over their kingdom from the top of a rocky outcrop, or kopje.
TOM & PAT LEESON

The spread of villages and farms fractures the ancient landscape, and wild creatures are marooned on islands of grass and trees surrounded by oceans of humanity. Unable to disperse over their former territories, populations of lions and other wildlife not only shrink but face the dangers of inbreeding. A natural example of the result has been studied by researchers in a population of lions isolated in the bottom of an ancient volcanic crater.

Ngorongoro Crater is a 600-meter (200-foot)-deep volcanic caldera at the eastern edge of the Serengeti Plain. A spectacular giant bowl with a floor of 260 square kilometers (100 square miles) and steep, high sides, this Garden of Eden holds an incredibly rich community of animals. Like birds in a gilded cage, the creatures of the crater, including about forty breeding lions, live in a lush land from which it is difficult to escape. Stranded with their small gene pool, the lions of Ngorongoro face an ever-shrinking choice of unrelated mating partners. The result is inbreeding, a suffocating trap that can be as fatal as bullets.

Painstaking detective work by lion researcher Craig Packer and his colleagues has reconstructed a fascinating history of the Ngorongoro lions during the past several decades. To track the relationships among members of the present lion population, they scouted the world for photographs of lions taken in the crater by scientists, tourists, professional photographers, and film crews who had visited the crater floor. Assembling and sorting their portfolio, they produced a who's who of individual lions that had lived in the crater over the years. Arranged chronologically, the record showed that the population had dwindled to only fifteen animals during the 1960s. The current crater lions, all descended from these individuals, show signs of inbreeding since the 1970s. They raise ever-fewer young, and studies reveal structural deformities in more than half the sperm of each male tested.

A pivotal event in the history of these lions was an outbreak of biting flies that occurred in the crater after a period of exceptionally heavy rains in 1961

FACING PAGE
The long tuft of dark hair at the end of a lion's tail is unique. Medieval observers believed that it concealed a horny spur or claw.

and 1962. Constantly breeding in the wet conditions, the fly population built up to intolerable levels. By mid-1962, most of the lions were so affected by festering sores and exhausted by lack of respite from the flies that they had become emaciated and unable to hunt. It is estimated that the resident population at this time fell to about eight. The survivors were joined by new males that must have entered the crater from the nearby plains, making up the fifteen lions from which later generations descended.

Lion numbers recovered to at least seventy by the mid-1970s, but, ironically, resident males now prevented further entry by new males that would have kept the gene pool more diverse and vigorous. With declining genetic diversity come infertility, birth defects, and weaker immune systems. In the long run, the damming of their gene flow and the fracturing of the species into small, inbreeding populations may be the greatest threat of all for lions and other wildlife at risk. Less diversity leaves them vulnerable, and a single germ might become an epidemic that wipes out an entire population.

The fate of the African lion seems set to follow the destiny of its cousin in India, where pressure from the human population has reduced this smaller species of forest lion to a population of only about 200 animals remaining in the world. The Asiatic lion was confined to the Gir Forest of western India as long ago as the start of the 20th century, protected by the nawab of Junagadh within his private hunting grounds. The forest covered about 2600 square kilometers (1000 square miles) at that time but has since shrunk to less than half that size and is now completely enclosed by cultivated land. The Gir Forest is the last block of natural vegetation in the semiarid region. Although ostensibly a wildlife sanctuary, it is home to about 7500 people and their 14,000 head of livestock. The last remaining lions are imprisoned in a small protected core of 259 square kilometers (100 square miles) designated as the Gir National Park. Within a 10-kilometer (6-mile) radius around the boundary of their sanctuary lives a human population of 160,000 and about 100,000 head of livestock.

It may finally be the cattle that sound the death knell for the lions. During drought years in the past, cattle have been moved by herders into the protected area to graze, and overgrazing and soil impaction from hundreds of thousands of hooves are slowly turning the forest into a desert. Exposed soils blow away, and forest plants find it difficult or impossible to become reestablished. Wild ungulates, the normal food of lions, feed mainly on woody plants rather than on grasses and do not damage the forest ecosystem; but their numbers, too, have been reduced by pressure from people and livestock.

Today, wild lions are restricted to the small remnant population in India and to southern and eastern Africa, with scattered populations ranging as far west as Mali and Senegal. The story of lions and people that began in prehistory is now entering its finale. With the inevitable arc of tragedy, the king of beasts has been hemmed into lilliputian domains by a species that presumes to rule over the entire world.

THE STORY OF LIONS AND PEOPLE THAT BEGAN IN PRE-HISTORY IS NOW ENTERING ITS FINALE.

The pressures of human population growth and the expansion of farmland in Africa pose a large question for the future of wild lions on the continent.

NOTES

Quotations in the text are from the following sources.

PAGE 20 From Schaller, *Golden Shadows, Flying Hooves,* quoting James Stevenson-Hamilton, a warden at Kruger National Park.

PAGE 65 From Schaller, *The Serengeti Lion,* quoting M. Edey.

PAGES 70–71 From Holman, *Inside Safari Hunting,* quoting Frederick Selous.

PAGE 72 From Roosevelt, *African Game Trails.*

PAGE 75 Ibid.

PAGE 78 "We are . . . surrounded," from Ritvo, *The Animal Estate.*

FOR FURTHER READING

BOOKS

Holman, Dennis, with Eric Rundgren. 1969. *Inside Safari Hunting*. London: W.H. Allen.

Owens, M., and Owens D. 1984. *Cry of the Kalahari*. Boston: Houghton Mifflin.

Packer, Craig. 1994. *Into Africa*. Chicago: Chicago University Press.

Ritvo, Harriet. 1987. *The Animal Estate: The English and Other Creatures in the Victorian Age*. Cambridge, Mass.: Harvard University Press.

Roosevelt, Theodore. 1910. *African Game Trails—An Account of the African Wanderings of an American Hunter-Naturalist*. London: J. Murray.

Schaller, George. B. 1972. *The Serengeti Lion*. Chicago: Chicago University Press.

Schaller, George B. 1973. *Golden Shadows, Flying Hooves*. New York: Alfred A. Knopf.

Zahavi, A., and Zahavi, A. 1997. *The Handicap Principle*. New York: Oxford University Press.

WEB SITES

www.lionresearch.org

The Lion Research Center under the direction of Dr. Craig Packer at the University of Minnesota includes descriptions of research projects, photographs, videos, maps, and opportunities for asking questions of the experts. You can also support research projects by making a tax-deductible donation.

www.lioncentral.com

This Web resource is dedicated to the lion and the other big cats, with information about the lion, image archives, reviews of books about lions and other big cats (including the most comprehensive listing of lion-related books on the Web), links to other lion Web sites, and lots of other material.

http://lynx.uio.no/catfolk/

The home page of the cat specialist group of the World Conservation Union leads to detailed information about all endangered cat species, including their biology, habitat, distribution, population, and protection status, principal threats to them, and references.

ACKNOWLEDGMENTS

The flavor of East Africa in these pages comes partly from my own experiences visiting there, but the details of lions and their nature draws on the published work of many authors, in particular the pioneering accounts of lions in the Serengeti by George B. Schaller. I was especially encouraged in my early research of the topic by conversations with Grant Hopcraft, a lion researcher in the Department of Zoology at the University of British Columbia. He also reviewed the manuscript and made many useful comments that helped improve it. I would like to thank Candace Savage for her comments on an early draft of the manuscript and Nancy Flight for editing and discussions throughout. Finally, I must acknowledge the forbearance of my young and energetic dog, Toby, who did his best not to ask me to play when I was sitting for long hours in front of the computer.

INDEX